What Is Sound?

Harcourt
SCHOOL PUBLISHERS

Orlando   Austin   New York   San Diego   Toronto   London

Visit *The Learning Site!*
www.harcourtschool.com

What Can You Hear?

Listen! What can you hear around you? You might hear birds chirping. You might hear someone singing or flipping the pages of a book. Be very quiet and listen to the world around you. What do you hear?

Outside you can hear many sounds. You might hear dogs barking. You might hear birds singing.

Fire trucks make loud noises. The noises alert people to move out of the way.

Sound is energy you can hear. All of the noises you can hear are caused by sound energy. A siren on a fire truck makes a loud sound. Loud sounds use a lot of sound energy.

Fast Fact

As something making a sound comes nearer to you, the sound gets louder.

Whispering is a very soft sound. Other sounds are so soft you can not hear them. Soft sounds take less sound energy to make.

CAUSE AND EFFECT What causes all the noises you hear?

3

How Are Sounds Made?

You hear sounds by listening with your ears. You can also make sounds. What are some ways you can make different sounds? You can make sounds by clapping your hands or drumming on your desk. You might whistle, hum, or blow on a horn.

Brass instruments make vibrations when you blow into them. The instruments then make sounds.

Vibrations made by hitting the top of a large drum make a deep booming sound.

All sounds are made when something vibrates. An object **vibrates** when it moves quickly back and forth. These back-and-forth movements are called vibrations.

Have you ever hummed into a kazoo? When you hum, a tiny object in the kazoo vibrates. It makes a buzzing sound. All sounds come from vibrations.

 CAUSE AND EFFECT What kinds of movements cause sounds?

5

Checking for Vibrations

Can you see objects vibrate? You might see the top of a drum vibrate. You might also see the edge of a cymbal or the strings of a guitar vibrate.

Stretch a rubber band over an empty box and pluck it. You can see the rubber band vibrate and hear the sound it makes.

Fast Fact

A sound becomes louder the more quickly an object vibrates.

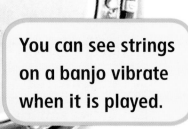

You can see strings on a banjo vibrate when it is played.

6

If you pinch a cymbal while it is ringing, the vibrations will stop.

Sometimes you can not see the vibrations. But you may be able to feel them. Hold your hand to your throat and hum a tune. You are feeling vibrations. When you stop humming, the vibrations stop. The sound they make stops, too.

 CAUSE AND EFFECT If you hit a drum top and made it vibrate, what would happen?

7

Waves of Sound

How does sound travel from a vibrating object to your ears? It moves in waves. **Sound waves** are vibrations that move through matter. The matter can be solids, liquids, or gases—such as air.

Gently tap a tuning fork. You hear a sound because the tuning fork vibrates. The vibrating tuning fork pushes and pulls on the air. It makes the air vibrate. You hear the sound when the vibrating air hits your eardrum.

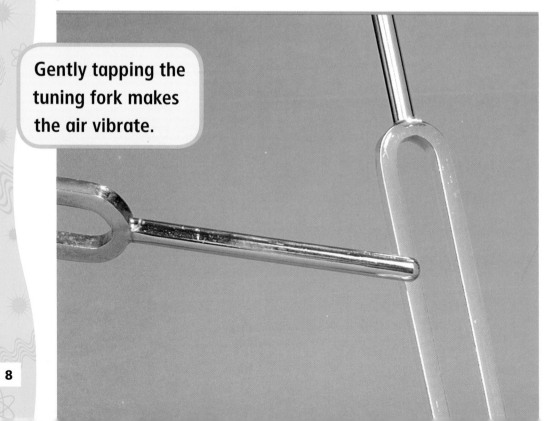

Gently tapping the tuning fork makes the air vibrate.

Gently tap the tuning fork again. Slowly lower it into a glass of water.

Watch what happens when the tuning fork touches the surface of the water.

The ripples you see in the water are sound waves moving through the water.

 MAIN IDEA AND DETAILS What is a sound wave?

The ripples you see in the water are sound waves moving through the water.

9

Did You Hear That?

Sound waves spread out in all directions. They can also travel a long way. You can hear bells from far away in any direction.

You know that sound waves move through solids, liquids, and gases. Sound waves move faster through liquid than through air. Sound waves travel fastest through most solids.

Sound waves move from the solid bells into the air. They are heard from all directions.

Dolphins use sound to communicate with one another.

Under water, dolphins "talk" to each other by making squeaks, clicks, and barks that move through the water. Humpback whales make sounds like songs that are carried through the water.

 MAIN IDEA AND DETAILS Do sound waves travel faster through water or through a wooden door?

Summary

Everything you hear is caused by sound energy. Sound energy is produced when objects vibrate. The vibrations make sounds. Sounds travel in waves through gases, liquids, and solids.

11

Glossary

sound Energy that you can hear. Sounds are made when an object vibrates. (3, 4, 5, 6, 7, 8, 9, 11)

sound wave Vibrations moving through matter. Sound waves can travel through solids, liquids, or gases. (8, 9, 10, 11)

vibrate To move back and forth quickly (5, 6, 7, 8, 11)